I0045558

Anonymous

The By-Laws of the Montreal General Hospital

As amended and finally passed by the governors and approved by the

corporation of the society of the Montreal General Hospital, on the 16th

May, 1876

Anonymous

The By-Laws of the Montreal General Hospital
As amended and finally passed by the governors and approved by the corporation of the society of the Montreal General Hospital, on the 16th May, 1876

ISBN/EAN: 9783337050252

Printed in Europe, USA, Canada, Australia, Japan

Cover: Foto ©berggeist007 / pixelio.de

More available books at **www.hansebooks.com**

THE

BY-LAWS

OF THE

𝕸𝖔𝖓𝖙𝖗𝖊𝖆𝖑 𝕲𝖊𝖓𝖊𝖗𝖆𝖑 𝕳𝖔𝖘𝖕𝖎𝖙𝖆𝖑

AS

AMENDED AND FINALLY PASSED

BY THE

**GOVERNORS AND APPROVED BY THE CORPO-
RATION OF THE SOCIETY OF THE MONTREAL
GENERAL HOSPITAL, ON THE 16th MAY, 1876,
TO WHICH IS PREFIXED A COPY OF THE
ORIGINAL AND OF THE AMENDED CHARTER.**

Montreal:
PRINTED BY LOVELL PRINTING AND PUBLISHING COMPANY,
St. Nicholas Street.
1876.

THE CHARTER,

Granted 30th January, 1823.

DALHOUSIE, *Governor.*

GEORGE the FOURTH, by the Grace of God of the United Kingdom of Great Britain and Ireland, King, Defender of the Faith.

To all to whom these Presents shall come, GREETING :

WHEREAS our loving subjects, John Richardson, William McGillivray, and Samuel Gerrard, Esquires, of our City of Montreal, by their humble Petition presented to our Right Trusty and Right Well Beloved Cousin, George, Earl of Dalhousie, our Captain-General and Governor in-chief in and over Our Province of Lower Canada, and read in Council for the said Province, on the ninth day of April, in the year of Our Lord one thousand eight hundred and twenty-two, did', among other things in substance set forth that there had been a subscription set on foot by them for the purpose of erecting a General Hospital in our said City of Montreal; and that sundry public spirited persons, influenced by principles of benevolence, have liberally subscribed towards the same; that, from the manifest utility of such an Hospital, further contributions and donations may be expected, if there be an assurance of permanency to the Institution, and means provided for the manage-

Petition of John Richardson, William McGillivray and Samuel Gerrard, Esquires.

ment of its concerns; that very considerable
progress has been made towards the object in con-
templation, by the purchase of a spacious lot of
ground in a central situation, in the Saint Lawrence
Suburbs of the said City, whereon is erected a large
building, for the body or centre part of the said
Hospital, and susceptible of extension hereafter
by wings upon a regular plan, which building
is now in a state of advancement towards com-
pletion, that promises eventual success if a Charter
of Incorporation be obtained; and, therefore, the
Petitioners humbly prayed for Our Letters Patent
forming a Corporation for the purposes aforesaid:

Which in
consideration
of its beneficial
tendency, is NOW WE taking into our Royal consideration, the
beneficial tendency of such an Institution, within our
said City, calculated for relieving the distresses of the
indigent, and preserving the lives of many useful
members of the community, are graciously pleased
to grant the said humble request of our said loving
subjects: KNOW YE, therefore, that We, of our
especial grace, certain knowledge, and mere motion,
granted. have willed, given, granted, ordained, constituted and
appointed, and, by these presents, for Us, our Heirs
and Successors, do will, give, grant, ordain, constitute,
and appoint, that Thomas Naters, John Molson, John
Richardson, William McGillivray, Samuel Gerrard,
John Forsyth, Frederick William Ermatinger, David
Ross, Thomas Phillips, Thomas Torrance, John Try,
George Garden, George Auldjo, Thomas Thain,
Robert Gillespie, Benaiah Gibb, John Molson the
younger, Thomas Molson, William Molson, Isaac
Winslow Clarke, William Hutchinson, Daniel Fisher,
Henry McKenzie, Charles William Grant, James
Millar, Archibald Norman McLeod, Alexander Skakel,
John Torrance, Joseph Chapman, Thomas McCord,
Angus Shaw, James Leslie, George Moffat, James

Reid, Robert Froste, William M. Porter, Charles
Brooke, Reverend John Bethune, David David, John
Fleming, Samuel Gale, Richard McGinnis, James
Stuart, Alexander Thain, Charles R. Ogden, Thomas
Busby, Abner Bagg, Charles F. Bancroft, John Blain,
Joseph Becket, William Caldwell, Henry Corse, Jacob
DeWitt, Jabez DeWit, Reverend Henry Esson, Hor-
atio Gates, Frederick Gonnerman, John Jones, Wil-
liam Kayes, Adam L. McNider, James B. Primo,
Andrew Shaw, Isaac Shay, John Wragg, James Wool-
rich, Zabdiel Thayer, William Stemm, James Dunn,
John G. Beek, Alexander Henry, Francis Metzler,
William Lunn, Charles Willard, Peter McGill, Ken-
neth Dowie, James Logan, Reverend James Somer-
ville, David Handyside, Joseph Shuter, Henry Loedel,
Levy B. Boruck, Hoyes Lloyd, Nahum Mower, Peter
McIntosh, Alexander Mabbut, Andrew Malcolm,
Robert McGinnis, John McKenzie, Joseph Nickless,
Cornelius Oakes, Jason C. Pierce, George Pyke,
Donald Robertson, Michael Scott, Stephen Sewel,
John Spragg, William Spragg, F. D. Strothers, Luther
Sawtell, Henry Solomon, John Simmons, Herman
Seavers, Ralph Taylor, John Jeffroy, Arthur Webster,
Albert Ware, George Wurtele, Oliver Wait, Andrew
White, Charles Wagener, Guy Warwick, Edward
Woolrich, Benjamin S. Solomon, T. R. Willard, Ken-
neth Walker, Andrew Allan, Robert Armour, William
Annesley, Thomas Blackwood, John Brown, James
Brown, Stanley Bagg, Abraham Bunbury, William
Binley, William Bent, J. T. Barrett, James Clark,
John Clarke, James E. Campbell, James Carsuel,
Benjamin Clamp, James A. Dwight, George Davies,
Peter W. Dease, Robert Drummond, John Fellows,
John Finley, Thomas French, Archibald Ferguson,
James Forster, William Forbes, James C. Grant,
William Gray, Christian C. Gunlack, Henry Griffin,

6

Nahum Hall, Benjamin Hart, Alexander Hardie, Robert Harwood, John Harwood, James Henry, Rukard Hurd, James Hughes, John P. Hogg, Charles Hoofstetter, Benjamin Hall, Isaac Jones, Nathaniel Jones, Nicholas Kurczyn, and all such other persons who have contributed, or shall hereafter contribute to the Institution, by a donation, respectively of five pounds current money of the Province of Lower Canada, or upwards of five pounds, and who shall regularly continue to pay annually towards its support not less than one pound of such money, shall be and become members of the Society and Corporation hereby erected, and forever hereafter shall, by virtue of these presents, be one Body Politic and Corporate, in deed, fact and name forever; and We will, give, grant and ordain, that they and their successors, "The Society of the Montreal General Hospital," by the same name shall and may have perpetual succession, and shall and may, by the same name, be persons capable in the law to sue and be sued, implead and be impleaded, answer and be answered unto, defend and be defended, in all Courts and elsewhere, in all manner of actions, suits, complaints, pleas, causes, matters and demand whatsoever, as fully and amply as any other of Our liege Subjects of Our Province of Lower Canada, may or can do by any lawful ways and means whatsoever; and that they and their successors, by the same name, shall forever hereafter be persons capable and able in the law to purchase, take, hold, receive and enjoy to them and their successors any messuages, tenements and real estate, whatsoever, and all other hereditaments of whatsoever quality they may be, in fee simple, for term of life or lives, or in any other manner howsoever; and also any goods, chattels and personal estate whatsoever; provided always, that the clear yearly value

What contribution shall constitute a Member.
And incorporated.
Name.
To have perpetual succession.
Capable to sue and be sued.
To hold estates,

of the said real estate doth not, at any time, exceed provided the clear yearly value does not exceed £1000 currency.
the sum of one thousand pounds lawful money of our
Province of Lower Canada, above all outgoings, and re-
prise, and that they and their successors, by the same
name, shall have full power and authority to give,
grant, sell, lease, demise and dispose of the said real To lease and sell estates.
estate and hereditaments whatsoever, for life or lives,
or years, or forever; and also all goods, chattels and
personal estate whatsoever, at their will and pleasure,
as they shall judge to be most beneficial and advanta-
geous for the good and charitable ends and purposes
above mentioned; and that it shall and may be lawful for them and their successors, forever hereafter,
to have a common Seal to serve for the causes and To have a seal.
business of them and their successors; and the same
seal to change, alter, break, and make new, from time
to time, at their will and pleasure; and our Royal
Will and pleasure is, that when our said Corporation
hereby erected, shall, by the general donations of
the benevolent, or otherwise, have sufficient funds, The Corporation shall complete the present building.
they shall complete the present building and offices,
erecting in the said Saint Lawrence Suburbs for the
said Hospital, and shall extend the same by wings,
or otherwise, so as to render the said General Hospital in all respects as perfect as may be, for the pur-
poses of the Institution which we Will shall forever which will be called "The Montreal General Hospital."
hereafter be called "The Montreal General Hospi-
tal;" and that it shall and may be lawful for our
said Corporation, from time to time, and at all times
hereafter, to erect, for their use and convenience, any
other house, houses or buildings whatsoever; but not
for other purposes than those of the said Hospital;
and for the better carrying into execution the purposes aforesaid, our Royal Will and pleasure is, and
We do hereby for Us, our Heirs and Successors, give
and grant to "The Society of the Montreal General

Hospital," and their Successors forever, that there shall be forever hereafter, belonging to our said Corporation, as many Governors for life of the said Hospital, as there shall be persons who have contributed, or shall contribute thereto, by donation respectively, twenty-five pounds, or upwards, lawful money of our said Province with an annual payment each of three pounds, or more, like money; and there shall be thirteen other Governors thereof, who shall be annually elected, in manner hereinafter prescribed, from among those persons who have contributed, or shall contribute to the Institution, under twenty-five pounds, money above said, and not less than ten pounds, with an annual payment thereto of two pounds, or more, like money, which persons so contributing, and paying, are hereby declared qualified to be elected Governors; which Governors for life, and those so elected shall appoint, out of their number, one President and one Vice-President; and also out of their number, or otherwise, one Treasurer and one Secretary, and shall conduct and manage the affairs and business of the said Hospital and Corporation for the ensuing year, in manner as hereafter is declared and appointed; and for the more immediately carrying into execution our Royal Will and pleasure herein, We do hereby assign, constitute and appoint the aforesaid Thomas Naters, John Molson, John Richardson, William McGillivray, Samuel Gerrard, John Forsyth, Frederick William Ermatinger, David Ross, Thomas Phillips, Thomas Torrance, John Try, George Garden, George Auldjo, Thomas Thain, Robert Gillespie, Benaiah Gibb, John Molson the younger, Thomas Molson, William Molson, Isaac W. Clark, Alexander Skakel, Daniel Fisher, Henry McKenzie, Charles W. Grant, Archibald N. McLeod, John Torrance, Joseph Chapman, Thomas McCord, James Reid,

Governors for life, those who contribute £25 or more, and pay annually £3 or more.

Governors elected, those who contribute less than £25, and not under £10, and pay annually £2, or more.

A President and Vice-President.
A Treasurer and Secretary.

First Governors of the Corporation.

9

Angus Shaw, James Leslie and George Moffatt to be
the present Governors of the said Hospital and Corporation, whereof the aforesaid John Richardson to be the
President, and the aforesaid John Molson to be the pre- First President,
sent Vice-President, the aforesaid Samuel Gerrard to Vice-President,
be the present Treasurer, and the aforesaid Alexander Secretary,
Skakel to be the present Secretary of our said Corpora- remain in office
tion hereby erected; which said President, Vice-Presi- until
dent, Governors, Treasurer and Secretary shall hold,
possess and enjoy their said respective offices, until the
first Tuesday in May next, ensuing, when the same the 1st Tuesday
shall cease and determine, as to this appointment; but in May, 1823.
continue, as to those aforesaid, who are created Governors for life; and for keeping up the succession in the And for keeping
said offices, our Royal Will and pleasure is, and We do, the Society is to
hereby, for Us, our Heirs, and Successors, establish, tion of Officers,
direct, and require of, and give and grant to the said Tuesday in May
Society of the Montreal General Hospital, and their annually,
Successors forever, that on the said first Tuesday in
May now next ensuing, and yearly, and every year forever thereafter, on the same day, (whereof public
notice of the hour, and place of meeting shall be given of which notice
in one or more of the Newspapers published in Mont- newspapers
real at least seven days before,) they and their success- seven days:
ors, the Members of the said Hospital and Corporation,
contributing and paying as aforesaid, shall meet at
the said Hospital, or at some other convenient place,
in our said City of Montreal, to be fixed, and ascer-
tained by some of the By-laws or Regulations of our
said Corporation, and there by the majority of such of
them as shall so meet, reckoning their votes in the the votes to be
ratio of the contributions actually paid; but so as that the contribu-
no member of the Society or Corporation shall have or otherwise; to
more than ten votes, shall by ballot, or in such other Governors.
manner and form as shall be directed by any of the
By-laws or Regulations of our said Corporation elect

The number of Governors, including those for life, not to be less than twenty-six. and choose thirteen of their members, qualified as
aforesaid, to be Governors, or whensoever the number

Out of the Governors, a President and Vice-President to be chosen. of Governors for life, then resident at Montreal, shall
be under thirteen, then such an addition shall be made
to the number so to be elected Governors, as to make
the whole number thereof (those so resident inclusive)
twenty-six, which Members of the Society and Corporation, so qualified and elected, shall be Governors

The Governors to enter upon their offices immediately, and remain in office for one year, or till others be chosen in their stead. of our said Hospital and Corporation for the ensuing
year, and with the President and Vice-President, by
them appointed, shall immediately enter upon their
respective offices and duties, and hold, exercise, and
enjoy the same respectively, from the time of such
elections and appointments, for and during the space
of one year, and until other fit persons shall be elected and appointed in their respective places according
to the Laws and Regulations aforesaid; and in case
any of the said persons so elected, and appointed to
the respective offices above said, or who shall be hereafter elected and appointed thereto, shall die, or be

In case of the death, removal, refusal, or neglect of officers others to be chosen in their stead by the Governors, removed from such offices respectively before the time
of their respective services shall be expired. or refuse
or neglect to act in and execute the office for which
he or they shall be so elected and appointed, then our
Royal will and pleasure is, and We do hereby, direct,
ordain and require, that the other Governors of our
said Corporation shall appoint a member or members
thereof, duly qualified, in the place and stead of him
or them so dying, removed or refusing or neglecting

within thirty days after such contingency. to act, within thirty days next after such contingency,
who shall serve until the first Tuesday in May next
following: and we do hereby will and direct. that this
method shall for ever after be used for filling up all
vacancies in the said offices between the annual elections above directed; and our will and pleasure is,
and we do hereby, for Us, our Heirs, and Successors,

direct, ordain and require that every President, Vice-President, Governor, Treasurer and Secretary of our said Corporation, to be elected and appointed by virtue of these presents, shall, before they act in their respective offices take an oath, to be to them administered by the President or Vice-President of our said Corporation for the time being or of the preceding year (who are hereby severally authorized to administer the same), for the faithful and due execution of their respective offices during their continuance in the same respectively; and further, our Royal will and pleasure is, and We do hereby for Us, our Heirs and Successors, ordain and appoint, and give and grant to the said Society of the Montreal General Hospital, that the President for the said Corporation for the time being and in case of a vacancy in the said office, or in case of his sickness or absence, the Vice-President shall, and may from time to time, as occasion may require, summon, and call together, at such places, within our said City of Montreal, as by any By-law shall be appointed for such meetings, and on such day and hours as the President or Vice-President shall respectively think proper, the Governors of the said Corporation and Hospital for the time being, giving them at the least one day's notice thereof; and We do hereby require them to meet accordingly, and give, grant, ordain that any seven, or more, of the Governors of our said Corporation, being so convened together, of whom the President, or in case of a vacancy in the said office, or the sickness or absence of the President, the Vice-President for the time being shall always be one, shall, for ever hereafter, be a legal meeting of the said Corporation, and they, or the major part of them so met, shall have full power and authority to adjourn from day to day, or for any other time, as the business of our said Corporation may require; and

Marginal notes:

Every officer, before he can act, must take an oath for the faithful discharge of his duties.

The President or in his absence the Vice-President, may summon the Governors to meet,

giving at least one day's notice.

Seven of the Governors, of whom the Vice-President or President shall constitute a quorum:

have power adjourn, and transact the business of the Corporation.

12

to do, execute, transact, manage, and perform in the
name of our said Corporation, all and every act and
thing whatsoever, which our said Corporation are, or
shall, by virtue of these our Letters Patent, be authorized to do, transact, manage, and perform, in as full
and ample manner, as if all and every the Governors
and Members of the said Corporation were present
and consenting thereto, saving and excepting always,
the electing of Governors, unless upon vacancies, as
aforesaid, happening in the intermediate period
between general elections; and also saving and excepting the giving, granting, selling, or otherwise aliening
any of the estate, real or personal, of our said Corporation, and the leasing, demising, or disposing of any
of the lands, tenements, hereditaments, real or mixed
estate, of our said Corporation, for any longer time
than one year; our Royal will and pleasure being,
that no part thereof be so sold, leased, or in any wise
aliened for any longer term or time, but by and with
the concurrence and approbation of the majority of
the whole number of the Governors of our said Corporation, for the time being, first obtained at any legal
meeting of the same; and further, We do, hereby, for
Us, our Heirs, and Successors, ordain and appoint-
and give, and grant, to the said "Society of the Montreal General Hospital," that at any such legal meeting of any seven or more of the Governors of our
said Corporation, of whom the President or Vice-
President, for the time being, shall always be one, it
shall and may be lawful for them, in writing under
the common Seal of our said Corporation to make,
establish, and ordain, from time to time, and at all
times hereafter, such By-Laws, Regulations, and Statutes, for the better government of the Officers, Members, and Servants of the said Corporation, and of
the Patients, from time to time, admitted into the

except the choosing of Governors, unless upon vacancies between general elections and except granting lands, &c., for a longer term than one year. None of the estate of the Corporation to be disposed of but by the consent of the majority of all the Governors.

The Governors in legal meeting may,

under the seal of the Corporation, make by-laws,

13

said Hospital; for fixing and ascertaining the place and for fixing the place and times of meeting of elections, and for regulating the manner, the of meeting of our said Corporation, on the days and times of election above mentioned; and for regulating the mode and manner of making such elections, the management and disposition of the funds and chari- management of the funds. ties, and all other the business and affairs of our said Corporation, as they, or the major part of them, so legally met shall judge best for the general good of the said Corporation, and profitable for promoting the charitable and beneficial designs of the said Corporation; and the same or any of them, to alter, To alter and repeal the by-laws, provided such laws be not contrary to the Charter, or the laws of the Province. amend, or repeal, from time to time, as they, or the major part of them, so met as aforesaid, shall judge most conducive to the benefit of the said charity; provided such Laws, Regulations, and Statutes, be not repugnant hereto, or to the laws of this our Province of Lower Canada; and We do further will, and grant that the said Governors for the time being, or any seven or more of them legally met as aforesaid, of whom the President, or Vice-President for the time shall always be one, shall have the full and sole power and authority for ever after, by the majority of their voices, from time to time, yearly, and every year, to To nominate Physicians, Surgeons, Apothecaries; nominate such and so many Physicians, Surgeons, and Apothecaries, as they shall judge necessary to attend to the said Hospital, and the sick and diseased patients, from time to time, admitted therein; and to appoint and designate the respective powers, authori- to ppoint their duties; ties, business, trusts, and attendances of the said Physicians, Surgeons and Apothecaries; and also to appoint a Steward, a Matron, a Nurse, or Nurses, and to appoint a Steward, and a Matron, and other servants, and their wages; all other servants and attendants upon the said Hospital, with their respective powers, authorities, business, trusts, and attendances, with the allowances that shall be found necessary from time to time, to be made and paid to any of the persons above said for their

14

attendance and services respectively in the said Hospital; and to displace and discharge any Steward,
Matron, Nurse, servant, and attendant, from the service thereof, and to nominate and appoint other, or
others, in their places or stead : and We do further
grant and ordain, that when, and as often, as any
President, Vice-President, Governor, Treasurer, Secretary, Physician, Surgeon, or Apothecary, of the
said Corporation, shall become unfit or incapable to
execute their said offices respectively, or shall mis-
demean themselves in their said offices respectively,
contrary to their duty, and to any of the By-Laws and
Regulations of our said Corporation, or refuse or neglect the execution thereof, and thereupon a charge
or complaint in writing shall be exhibited against
him, or them, by any member of our said Corporation,
at any legal meeting of the Governors thereof, as
aforesaid, that it shall, and may be lawful for the
President, or Vice-President and Governors, or the
major part of them then met, or at any other Legal
meeting of the said Corporation, from time to time,
and upon examination and sufficient proof, to suspend
or discharge such President, Vice-President, Governor,
Treasurer, Secretary, Physician, Surgeon, or Apothecary, from their offices respectively, although the
yearly, or other time of their respective services shall
not be expired, anything in these presents before
contained to the contrary in any wise notwithstanding; provided always, that none of the said Officers
so complained against be suspended or discharged at
any meeting, without the concurrence and approbation of the majority of the whole number of the
Governors of the said Corporation, nor without having
a copy of the complaint or charge against him at least
six days before such examination, and an oppor-
tunity to be fully heard in his defence. And our will

To displace the
Steward, Ma-
tron, or other
servant.

The President,
Vice-President,
Governor,
Treasurer,
Secretary,
Physician,
Surgeon, or
Apothecary
becoming unfit
or misdemeaning
himself,

may, upon
sufficient proof,
be discharged
from his office,

but not without
the consent of a
majority of all
the Governors,
and having a
copy of the
charge six days,
and being heard
in his defence.

15

and pleasure further is, that the said Governors of the
said Corporation and Hospital shall, from time to
time, when thereunto required by the Governor, Lieutenant-Governor, or Person administering the Government of our Province of Lower Canada, give an account, in writing, of the several sums of money by
them received and expended, by virtue of these presents or any authority hereby given, and of the
management, application, and disposition of the re-
venues, donations and charities aforesaid, to such
person or persons as the said Governor, Lieutenant-
Governor, or Person Administering the Government,
shall from time to time, appoint to receive and audit
the said accounts: And further, We do, by these
presents, for Us, our Heirs or Successors, give and
grant unto the said "Society of the Montreal General
Hospital," and their Successors for ever, that this our
present Charter shall be deemed, adjudged and construed, in all cases, favorably, and for the benefit and
advantage of our said Corporation, and for promoting
the good of this charitable Institution; and this our
present grant being entered on Record, as is hereinafter expressed, shall be for ever hereafter, good and
effectual in the Law, according to our Royal intent
and meaning hereinbefore declared, and without any
other license, grant or information from Us, our Heirs
or Successors, hereinafter by the said Corporation to
be had or obtained, notwithstanding any misrecitals,
not naming or misnaming, of any of the aforesaid
Offices, Franchises, Privileges, Immunities, or other
the Premises, or any of them; and although no writ
of *ad quod damnum*, or other writs, inquisitions, or
precepts, has been, upon this occasion, had, made,
issued or prosecuted, any statute act, ordinance
or provision, or other matter and thing, to the
contrary thereof notwithstanding: provided further,

Margin notes:
When required by the Governor, &c., of the Province, the Governors are to render an account of the moneys received and expended.

The Charter to be construed in favor of the Society;

and on being entered on Record, shall be effectual in law.

and Our will and pleasure is, that the said Society and Corporation of the Montreal General Hospital as hereby constituted, made, erected and created with the express reservation and condition that our Letters Patent relating thereunto may be hereafter altered and amended in such manner and form, and with such additions, diminutions, extensions and explanations as our Governor, Lieutenant-Governor, or Person administering the Government of our Province of Lower Canada, by and with the advice and co..sent of our Executive Council of our said Province, shall judge best for the general good of the said Corporation, and for the more effectually promoting the charitable and beneficial designs of the said Society, according to the true intent and meaning of the contributors thereto.

May be altered and amended, as the Governor, &c., with the advice of the Executive Council, shall judge for the good of the Society.

In testimony whereof, We have caused these our Letters to be made Patent, and the Great Seal of our Province to be hereunto affixed, and the same to be entered of record in our Secretary's Office for our said Province of Lower Canada, in one of the books of Patents there remaining.

Witness our Right Trusty, and Well Beloved Cousin, George, Earl of Dalhousie, our Captain-General and Governor-in Chief in and over our said Province of Lower Canada, at our Castle of St. Lewis, in our City of Quebec, by and with the advice and consent of our Executive Council for our said Province of Lower Canada, the thirtieth day of January, in the year of Our Lord one thousand eight hundred and twenty-three, and of our Reign the Fourth.

(Signed,) Ls. MONTIZAMBERT,   D.,
            Acting Provincial Sec.   G.

THE
AMENDED CHARTER.

22 VICT. CAP. 116.

" An Act to amend the Charter of the Society of the Montreal General Hospital."
Assented to 4th May, 1859.

WHEREAS the Society of the Montreal General Preamble.
Charter of 30th
January, 1823
Hospital by their petition to the Legisla-
ture, have set forth, that on the thirtieth day of
January, in the year of our Lord one thousand eight
hundred and twenty-three, they were duly constituted
a body politic and corporate, under and by virtue of
certain Letters Patent of His late Majesty King George
the Fourth, duly issued on that day at the Castle
of St. Lewis in the City of Quebec, under the great
seal of the then Province of Lower Canada, and have
ever since acted and still act as such thereunder; that
certain of the provisions of the said Letters Patent,
more especially in reference to the qualification of the
members of the said Corporation ;—its powers in
respect of the holding and alienating of property,
—the number, choice, and qualification of the Governors thereof,—the quorum of Governors for the
transaction of business and the extent of their powers
of administration, are found in practice to be highly
inconvenient ; and that they therefore pray for amendment of their said Charter; and whereas it is expedient to grant their said prayer : Therefore Her

D

Majesty, by and with the advice and consent of the
Legislative Council and Assembly of Canada, enacts
as follows:

Who shall be members of the Corporation. 1. The Present Gove.   ~s of the said Hospital, and
all other persons who, du.   the year to expire on the
first Tuesday of May, in the year one thousand eight
hundred and fifty-nine, shall have contributed to its
funds five dollars or more, are hereby declared to be
members of the said Corporation: and hereafter, all
persons who, during the current or last expired financial year thereof, shall have contributed to its funds
five dollars or more shall be such members thereof,—

Provided always, that no person shall be entitled
to vote as such member, who shall not have actually
paid up such contribution for the then current
financial year.

What real estate the Corporation may acquire and hold. 2 The said Corporation may acquire and hold, by
any legal title whatsoever, real estate of a yearly
value not exceeding four thousand dollars, as by the
said Charter allowed, besides what they may require
for the actual occupation of their Hospital; and they
may acquire any other real estate, or interest therein,
by gift, devise or bequest, if made six months or
upwards before the death of the party making the
same, and may hold the same for a period of not more
than five years; but the same, or any part thereof, or
interest therein, which may not within the said period
have been alienated, shall revert to the party from
whom the same was acquired, his heirs or other
representatives.

Who shall be life Governors. 3. The present life Governors of the said Hospital
shall continue so to be, subject only to the condition
of their continuing to contribute to its funds the
yearly sum of twelve dollars or more, as hereto-
fore,—but may be declared by vote of the Board of
Governors to have ceased so to be, if at any time two

years in arrear for such contribution; and any other persons who shall have contributed by donation to its funds one hundred dollars or more, may be constituted such like Governors by vote of the said Board, subject to the like condition of contributing such yearly sum of twelve dollars or more,—and may in like manner, be declared by vote of the said Board to have ceased so to be, if at any time two years in arrear for such further contribution.

4. The present elected Governors of the said Hospital shall also continue so to be until the time of the next annual election of Governors, to be held under the said Charter, after the passing of this Act, and shall severally be eligible for re-election thereat, if otherwise qualified; and at such next annual election there shall be chosen by the members of the Corporation, qualified to vote, such number of elected Governors not exceeding six as may be deemed expedient, to serve for the term of one year and until their successors shall be elected, and a like further number to serve for the term of two years and until their successors shall be elected; and at every annual election thereafter, a like number of not more than six elected Governors shall in like manner be chosen for the like term of two years and until their successors shall be elected to supply the place of those retiring, who, however, shall always be eligible for re-election, if otherwise qualified; and all persons, donors of forty dollars or more, if also contributors of eight dollars or more yearly, to the Funds of the said Hospital, and also all contributors of twelve dollars or more yearly to its funds, shall be held qualified for election as such Governors.

Elected Governors:—

5. In case of the death, resignation or disqualification of any elected Governor, it shall be in the discretion of the Board of Governors to name another

Vacancies among elected Governors.

qualified member of the said Corporation, to serve in his stead for the unexpired remainder of the term for which he was elected.

President and Vice-President, election, of term of Office, &c.

6. There shall continue to be elected by the Board of Governors, from among themselves, as soon as conveniently may be after each annual election of Governors, a President, and Vice-President of the said Corporation, who shall have such powers and discharge such duties as by By-law in that behalf may be ordained, and shall serve for the term of one year and until their successors shall be elected; and in case

Vacancies in the office.

of the death, resignation, or disqualification of such President or Vice-President, the said Board, as soon as conveniently may be thereafter, shall elect another of themselves to serve as such for the unexpired re-

Governors may appoint and remove officers and servants of the Corporation, &c.

mainder of his term of office; and the said Board may, further as occasion shall require, appoint all such other officers, and employ all such servants of every grade, of the said Corporation, on such terms, in respect of duty, emolument and otherwise, as by By-law in that behalf may be ordained, and may remove all such officers and all such servants, in their discretion, subject only to such restrictions as by By-law in that behalf may be ordained.

Board of Governors to administer the affairs of the Corporation

7. The said Board of Governors shall in all other respect have full power to administer the affairs of the said Corporation, subject only to such restrictions as its By-laws may ordain, and in particular, may sell or in any other way dispose of any estate, real or personal of the said Corporation, as they may deem advisable for the interests of the said Corporation;

Proviso; as to investment of property sold.

Provided always, that all moneys from time to time to be received from them on account of purchase money of any real estate by them alienated or to be alienated, or on account of the capital of any ground rent, or otherwise than by way of contribution not made for investment, shall be dealt with as capital only, and not as income, and shall be promptly invested either in buildings, or other real estate for the occupation of the said Hospital, or in productive real estate, or upon security thereof, or in public securities of the Province.

8. So much of the said Charter as provides that every President, Vice-President, Governor, Treasurer, and Secretary of the said Corporation shall take an oath of office, is hereby repealed. *President, &c., not to take an oath of office.*

9. The Quorum of the Board of Governors for the transaction of all business, is hereby reduced to *five*; and the presence of the President or Vice-President shall not be necessary to constitute such Quorum. *Quorum of Governors fixe.*

10. The present By-laws of the said Corporation, in so far as they may not be contrary to any provision of the said Charter as hereby amended, or to law, shall remain in force until duly repealed or amended. *Present By-laws continued.*

11. The said Board of Governors shall have power hereafter to provide by By-law, from time to time, for any changes which may be deemed expedient as to the time prescribed by the said Charter for holding the annual meetings of the said Corporation, or as to the notice thereby required to be given for such meetings, or as to the rule of voting at such meetings thereby prescribed, or as to the mode therein indicated for the summoning of meetings of the said Board of Governors, as also for any lowering of the qualification hereinbefore set forth of elected Governors, which may be deemed expedient, or for the requiring of any larger Quorum of the said Board of Governors than is hereinbefore set forth, whether for transaction of business generally or of any particular description of business, as may be deemed expedient, and generally for all other matters and things whatsoever *Board of Governors may make By-laws for certain purposes, and may repeal or amend them.*

appertaining to the affairs of the said Corporation; and they may repeal or amend all such By-laws:

Proviso. Provided always, that no By-law, and no repeal or amendment of any By-law, shall take effect until after approval thereof by vote of the members of the said Corporation, at an Annual or Special Meeting of the Corporation duly called.

Corporation to make returns to the Legislature. 12. The said Corporation shall at all times, when thereunto required by the Governor or by either branch of the Legislature, make a full return of its property, real and personal, and of its receipts and expenditure, for such period and with such details and other information, as the Governor or either branch of the Legislature may require.

Public Act. 13. This Act shall be deemed a Public Act.

THE
BY-LAWS

OF THE

Montreal General Hospital,

AS

AMENDED AND FINALLY PASSED BY THE GOVERNORS AND
APPROVED BY THE CORPORATION OF THE SOCIETY OF THE
MONTREAL GENERAL HOSPITAL, ON THE 16th MAY, 1876.

CHAPTER I.

OF THE ELECTION OF GOVERNORS.

1. Notice of the hour and place of meeting, for the election
of Governors, shall be given by the Secretary in one or more
of the newspapers published in Montreal, at least seven days
previous to such meeting.

2. The votes shall be given in the ratio of the annual
contributions ; that is, one vote for every five dollars contributed. The Life and elected governors shall have five votes
in addition to those they are entitled to by their annual
contribution, but in no case shall the votes exceed ten.

3. The annual meeting of the Corporation for the election
of Governors and other business shall be held on the third
Tuesday of May, in each and every year, in the Governor's
Room of the Hospital, at three o'clock P.M., and continued
until four o'clock P.M. of the same day; at which two
Governors for the preceding year shall be appointed to
inspect the ballots; and every person voting at said election

shall personally and then and there appear and deliver to the aforesaid Inspectors a ballot containing the names of six persons, as Governors for the ensuing two years; and the Inspectors shall deposit in a bag all ballots so delivered, and shall insert the names of the persons so voting in a poll-list kept by them for the purpose: and when the poll of such election shall be closed, the Inspectors shall open and count the said ballots, and shall openly declare to the meeting the names of the persons who shall be found to have been elected by the majority of the votes, and shall deliver a certificate thereof to the Secretary, who shall forthwith communicate the same to the persons so elected.

4. The Secretary shall immediately after the election, give notice in writing to the Governors elected, and to the Governors for life, requiring them to meet on the Thursday next ensuing at 3 o'clock P.M. to choose a President, a Vice-President, a Secretary. a Treasurer, a Committee of Management, and Medical Officers, for the ensuing year.

CHAPTER II.

OF THE GOVERNORS.

1. A quarterly meeting of the Governors shall be held in the Governor's Room of the Hospital, or such other place as may be appointed by the Governors, on the second Wednesday in August; on the second Wednesday in November; on the second Wednesday in February; and on the second Wednesday in May,—in each and every year, at three o'clock P.M.

2. The Governors for life, together with those elected, shall choose by ballot, out of their number, on the first Thursday after the annual meeting, in each and every year, one President and one Vice-President; and, out of their number or otherwise, one Secretary and one Treasurer; and also, out of their number, they shall nominate and appoint

by ballot, a Committee of Management, consisting of ten in
addition to the President, Vice-President and Treasurer, who
shall be members ex officio. Three members of the Committee of Management shall form a quorum.

3. The Governors, at their meeting on the first Thursday
after the annual meeting in each and every year, shall elect
(by vote and not by ballot) eight Medical Officers, to be the attending Physicians and Surgeons for the year, and four out-
door Physicians and Surgeons, to attend to the out-door
patients, and a specialist to act as oculist and aurist. No
person shall be eligible as a member of the Medical Board
unless he shall have been nominated by a governor and
seconded by another governor; such nomination to be posted
in the Governor's Hall at least two whole days before the day
of election.

4. If any vacancy occur in the Medical Board it shall be
filled up at the ensuing quarterly meeting of the Governors,
or at a special meeting of the Governors called for that
purpose, after ten days' notice shall have been give to the
Governors in the usual way.

5. The President, or in his absence the Vice-President
may, or at the request of seven Governors, shall summon
and call together in the usual way the Governors of the
Corporation for the time being; giving them (except in
the case specified in the foregoing rule) at least three
days' notice thereof, and intimating the purposes of the
meeting.

6. Five or more Governors, convened according to the
By-Laws of the Corporation, shall constitute a legal meeting
of said Corporation, and shall be competent to transact all
business relative to the said Corporation, in as far as they
are authorized by the charter.

7. Two of the Governors shall, in rotation, visit the Hospital at least twice a week; and shall enter their names in a
book, with such observations relating to the Institution as
they shall think necessary.

CHAPTER III.

OF THE PRESIDENT AND VICE-PRESIDENT.

1. The President, or in his absence the Vice-President, shall preside at all meetings of the Governors of the Corporation ; and in the absence of both the President and Vice-President. a Chairman *pro tempore* shall be appointed by the meeting.

2. At all meetings of the Governors, every person shall, in speaking, address himself to the Chairman, who shall propose all questions; and if a difference of opinion shall arise on any question, such question shall be fairly stated by the Chairman, and shall than be determined by ballot (if two of the Governors require it); and if in any case the votes be equal, the Chairman shall have the casting vote.

3. In selling or acquiring real Estate, and in accepting and discharging mortgages on real Estate, the President, or in his absence the Vice-President, the Treasurer and the Secretary shall together act in behalf of the Corporation as its legal representatives.

4. The President, or in his absence the Vice-President, together with the Treasurer shall have custody of all bonds and securities, title-deeds and other papers relating to the property of the Corporation.

CHAPTER IV.

OF THE TREASURER.

1 The Treasurer shall give such security as may be required by the Governors for the due accounting for and payment of the monies which he shall from time to time receive for the use of the Hospital.

2. He shall also receive all monies belonging to the Corporation, and deposit the same in one of the chartered banks

of the city; and all payments shall be made by cheque to be signed by the Chairman of the Committee of Management, and countersigned by the Treasurer.

3. He shall deliver to the Secretary, three days at least previous to the General Annual Meeting in each and every year, a full statement of all his accounts signed by the Auditors, to be laid before the said meeting.

CHAPTER V.

OF THE SECRETARY.

1. The Secretary shall attend at all meetings of the Governors, take minutes of the proceedings of each meeting, and transcribe the same into a book.

2. He shall at each meeting of the Governors read the proceedings of the last meeting.

3. Through him all communications, of whatsoever nature, shall be made to the Governors, and by him from the said Governors to the public, or any department of the Hospital, except in respect of such communications as may be more properly signed by the President.

4. He shall, immediately after the election of the Governors or other officers of the Hospital, give notice thereof in writing to the different persons so elected.

5. At least one day previous to each and every meeting of the Governors, it shall be his duty to send notice of the time and place of such meeting to each Governor, stating the purpose of the meeting.

6. He shall, seven days before any election of Governors, hang up a notice in the Committee Room of the Hospital, and shall state therein the names of the persons qualified to be elected, as well as those who are qualified to be the electors: those qualified being in the former case, donors of forty dollars or more and annual contributors of eight dollars or

more, or annual contributors of twelve dollars and upwards, and in the latter case contributors of five dollars and upwards for the last expired financial year.

7. He shall keep the necessary books for entering all the transactions of the Governors, and furnish such reports as they may require.

8. The Secretary shall lay on the table, in the Committee Room, at every meeting of the Governors, the journal or minutes of their proceedings; the case-books of the Medical Board; the book containing the minutes of the visiting Governors; and any other document relating to the affairs of the Corporation which they may require.

CHAPTER VI.

OF THE COMMITTEE OF MANAGEMENT.

1. They shall, at their first meeting after their appointment, elect, by vote, a Chairman and a Secretary.

2. The Committee of Management shall meet at the Hospital at least once a week; and shall have power, and it shall be their duty, to inquire strictly into the domestic economy of the house; to order all things necessary for the same; to inquire into the conduct of the Officers, Matrons, Stewards, and Servants, of the Hospital; to order all repairs; to oversee and control everything connected with the management of the Hospital; to audit, inspect and approve all accounts, and to order payment of the same; and to report, through the Secretary of the Corporation, a statement of all their transactions, to each quarterly meeting of the Governors.

3. They shall have the general charge and care of the property of the Corporation, and shall enforce all necessary attention to economy, cleanliness and good conduct.

4. They shall at their meetings make minutes of their

proceedings, which they shall forthwith cause their Secretary to transcribe into a book.

5. They shall have the power and it shall be their duty to engage and discharge and prescribe and regulate the duties of all Servants of the Hospital, and fix their wages.

6. The Committee of Management shall, at each quarterly meeting of the Governors, report the state of the Hospital, and the state and number of patients admitted and discharged during the preceding quarter.

7. They shall insert or cause to be inserted in a book, to be provided for the purpose, all orders to the Steward for the purchase of the stores, which orders must be signed by a member of the Committee.

8. They shall lay on the table, at every meeting of the Governors, the book containing the minutes of the proceedings.

9. When a vacancy occurs in the Committee of Management the remaining members of the Committee may elect from the Board of Governors a person to fill the vacancy for the unexpired term.

CHAPTER VII.

OF THE MEDICAL BOARD.

1. The Medical Board of the Montreal General Hospital shall consist of the consulting and attending Physicians and Surgeons, and of the out-door Physicians and Surgeons.

2. The number of the attending Medical Officers shall be eight, who shall make such arrangements among themselves that the Hospital be visited once every day at least by two of them, and there shall be four out-door Physicians and Surgeons, one or more of whom shall attend daily and prescribe for the out-door patients.

3. No person shall be eligible to the office of attending

Physician or Surgeon, or of out-door Physician or Surgeon, who is not a graduate of a University or College in the British Dominions, or a licentiate or member of a College of Physicians or Surgeons of Great Britain and Ireland, or other in the British Dominions, of at least three years standing, and who is, moreover, not licensed to practice in this Province.

4. Any attending Physician or Surgeon may, after having acted in that capacity for twelve years, become a consulting Physician or Surgeon of the Institution.

5. The visiting hours for the Medical Officers attending the wards shall be twelve o'clock (noon) of each and every day, Sundays excepted.

6. Should any one of the attending Physicians and Surgeons, or of the out-door Physicians and Surgeons, be unable, from illness or intended absence from town, to fulfil the duties of his office, such attending Physician and Surgeon shall obtain the assistance of some other of the attending Physicians and Surgeons for the time being, and such out-door Physician and Surgeon shall obtain the assistance of some other of the out-door Physicians and Surgeons for the time being.

7. In all surgical cases requiring operations (except in emergencies where delay might be dangerous), and in medical cases when desired, a consultation of the whole Medical Board of the Hospital shall be summoned.

8. The Medical Board shall have full power, at all times, to visit and examine the Apothecary's shop.

9. It shall be the duty of the Medical Board to examine all Candidates for the situation of House Surgeon, assistant House Surgeon and Apothecary.

10. The Medical Board shall communicate to the Committee of Management, in writing, through their Secretary their opinion of the candidates.

11. The attending Physicians and Surgeons shall alone have the power to admit and discharge any patient, except in

cases of severe injuries, where the House Surgeon shall admit at any hour of the day or night, without recommendatory letters.

12. The Medical Board shall regulate the diet of the patients.

13. The attending Physicians and Surgeons and the outdoor Physicians and Surgeons shall have the power to name and appoint their own Dressers and Clerks from among the students.

14. For the purpose of Pathological investigation, the attending Medical Officers of the Hospital shall have the power of examining, post mortem, the bodies of patients deceased within it, at their discretion; and such examination shall only be made in the presence or by the order of the same.

15. The Medical Board shall make suitable arrangements amongst themselves concerning the service of the out-door department.

16. All minor surgical operations not requiring admission to the wards may be performed by the out-door Physicians and Surgeons; when admission to a ward is required by the patient or case, it will be the duty of the attending Physician and Surgeon to take charge of the patient and perform the necessary operations.

CHAPTER VIII.

OF THE OCULIST AND AURIST.

1. The Oculist and Aurist shall attend the out-door room on such days and at such times as shall be determined by the Medical Board.

2. He shall have the right to admit to two beds in the male public and two in the female public wards, patients suffering from ophthalmic or aural complaints.

3. He shall have the privilege of admitting to one private

ward when the same happens to be vacant, but he shall at no time claim the use of more than one private ward.

4. He shall keep a register and case-book, including all cases coming under his department.

5. He shall not be *ex-officio* a member of the Medical Board.

CHAPTER IX.

OF THE RESIDENT MEDICAL OFFICERS.

1. No person shall be eligible as House Surgeon or Assistant House Surgeon unless he be a Graduate or Licentiate of some University or College in the British Dominions and a Licentiate of this Province; and, further, unless he submit to the Committee of Management satisfactory testimonials of his moral character, and of his having been approved of by the Medical Board, and having attained the age of 21 years.

2. The House Surgeon shall visit all the occupied wards in the Hospital at least once every morning and evening, and oftener when required so to do by the attending Medical Officer, and shall examine all cases of serious disease or accident admitted after the visit of the attending Medical Officers, and take charge thereof until the next visit of those officers; he shall also be prepared to report to the attending Medical Officers the state of the patients since the last visit.

3. In any case of an urgent nature, the House Surgeon shall immediately summon the attending Medical Officer; and if he cannot be found, he shall call upon the nearest member of the Medical Board.

4. The House Surgeon and Assistant House Surgeon shall not absent themselves from the Hospital during the visits of the attending Medical Officers; and when at any time either goes out, he must notify his colleague, so that either the House Surgeon or Assistant House Surgeon may be always present in the Hospital.

5. They must always be in the Hospital at a seasonable hour in the evening; and never sleep out of the building unless with the permission of one of the attending Physicians.

6. The House Surgeon and Assistant House Surgeon shall record in the case-books, to be kept for that purpose, the history, symptoms, treatment, and results of treatment, of all patients admitted to their respective wards, as well as the details of the morbid appearances revealed by the post mortem examinations conducted by the attending physicians.

7. The House Surgeon and Assistant House Surgeon may, and in the absence of the apothecary shall personally supervise the compounding of all prescriptions ordered for patients.

8. The House Surgeon shall have charge of all the surgical instruments belonging to the Hospital, and shall be responsible for the same; and he shall enter in the register of patients the name of the person by whom the patient has been recommended, and whether an ordinary or a pay-patient, and the rate per week; and the time and the manner in which every patient leaves the Hospital.

9. It shall be the House Surgeon's duty to notify the members of the Medical Board whenever a consultation is required.

10. The House Surgeon shall draw up a quarterly and an annual report of the state and number of patients admitted and discharged previously to the quarterly meetings of the Governors, and the Committee of Management, and the annual meeting of the Corporation.

11. The House Surgeon shall notify the Steward of the dismissal of all patients.

12. In all cases when it becomes necessary to notify the Coroner of the sudden death of any individual within the precincts of the Hospital, it shall be the duty of the House Surgeon to give such notice; and on the holding of the inquest, he shall be the Medical Officer referred to for information by that functionary and the Jury summoned by him and shall be entitled to the fees paid by the Coroner in such cases.

13. The Assistant House Surgeon shall take charge of such wards as are assigned to him by the attending Medical Officer, and shall visit these at least once every morning and evening, and oftener when required so to do by the attending Medical Officer; and in the event of his observing any great or sudden change in any of the patients in his wards, he shall immediately notify the House Surgeon to that effect.

14. Neither the House Surgeon nor his assistants shall be at liberty to take outside practice; but must devote their time wholly to the interests of the Institution.

CHAPTER X.

OF THE APOTHECARY.

1. No person shall be eligible as Apothecary until he shall have laid before the Committee of Management satisfactory proofs of moral character, and of his having been examined and approved of by the Medical Board. He shall reside in the Hospital.

2. He shall compound, make up, and dispense all the medicines required for the purposes of the Institution.

3. He shall regularly attend the apothecary shop, of which he shall have charge; keep everything belonging thereto clean and in proper order; and never absent himself from the Hospital without the knowledge of the House Surgeon.

4. He shall deliver to the nurses of the wards the different medicines, &c., ordered for the patients of each; the directions for which shall be legibly written or printed, and affixed to each phial, box, or parcel.

5. He shall be permitted to instruct in pharmacy the pupils attending the Hospital.

CHAPTER XI.

OF THE STUDENTS ATTENDING THE HOSPITAL.

1. No Student who has not previously taken a ticket for attendance in the Hospital shall, on any pretence whatever, be allowed to witness any operation or dissection, or to attend the Clinical Lectures.

2. The fee for a Student's annual ticket shall be....$12.00
Six months.. 8.00
Perpetual if paid in one sum on entrance........ 20.00
If otherwise... 24.00

3. Students following the attending Medical Officers during the visits are to behave with decency and propriety, keeping their hats off at all times; and they must avoid doing anything that may tend to disturb any official or patient.

4. Every Student must keep off his hat while he is in the operating theatre, both that he may not obstruct the view of others and as a mark of respect; and all noises and changes of seats must be avoided, as unpleasant to the operator and hurtful to the patient.

5. Students when visiting the wards are on no account to offer any advice, or express any opinions, unless when requested so to do by the attending Medical Officer.

6. No Student shall enter the wards at any time except the visiting hour, unless by permission of the attending Medical Officer.

7. Any Student infringing any of the above regulations shall forfeit the benefit of his ticket and the privilege of attending the Hospital.

CHAPTER XII.

OF THE ADMISSION OF PATIENTS.

1. Every Governor and Member of the Medical Board of the Hospital, may recommend patients to be admitted into

the Hospital in the manner and form hereinafter prescribed; but it shall be the exclusive prerogative of the attending Physician or Surgeon, as the case may be, to judge if the person so recommended be laboring under such disease as is admissible into the Hospital, according to the By-Laws of the Corporation. Provided that should any patient recommended by a Governor be refused admission, the Medical Officer so refusing shall record his reasons for so doing with the Steward, who shall communicate the same to the Governor by whom the applicant was recommended.

2. In any urgent case, when delay may be dangerous, the House Physician or Surgeon shall have power to admit patients; and he shall immediately give information of the same to the Attending Physician or Surgeon, as the case may be, or, if he cannot be found, to any other of the Medical Board of the Hospital.

3. No patient shall be admitted into the Hospital whose case shall be considered incurable, or who is insane, or whose case does not require the particular benefit of an in-door practice.

4. No woman in advanced pregnancy shall be admitted as an in-door patient unless she is the subject of serious accident or some disease dangerous to her life, and requiring immediate care and treatment.

5. No children under two years of age, except in cases requiring special treatment, shall be admitted into the Hospital.

6. Such patients as are unable to pay for their maintenance shall be received as " ordinary " patients. Such as are able to pay, shall be admitted into the Hospital as " pay " patients, and none of these at less than one dollar and a half per week; for the payment of which the person recommending such patient shall be responsible.

7. The person recommending any patient shall mention,

in the certificate for admission, if the patient be an
" ordinary " or " pay " patient.

8. No patient shall be admitted into the Hospital as an
" ordinary " patient, excepting immigrants, who does not
reside in the city or parish of Montreal ; and the person or
persons who shall recommend any patient who does not
reside in the city or parish aforesaid, shall become responsible for the payment of the Hospital dues of such patient,
which shall not be less than one dollar and a half per week ;
and in case of his or her death, he shall also be bound to
defray the expenses of the funeral, which shall not exceed
five dollars. But, as special cases may occur when a devia-
tion from this By-Law may be necessary, the Committee of
Management are authorized to dispense with it in such
cases, and shall report the same to the Governors at their
next meeting.

9. All persons the subject of venereal disease, applying for
admission shall be called " Lock-ward " patients, and shall
pay at least two dollars per week, and if it be not specified
on the ticket of admission that he or she is a Lock-ward
patient, then the attending Medical Officer shall send back
such patients to the person recommending, with the words
" Lock-ward " written on the ticket; and if the person
recommending shall renew his recommendation, he shall then
be responsible for the payment of the Hospital dues for such
patients. All female Lock-ward patients shall be admitted
only into a ward devoted to such cases, to be called the
" Lock-ward."

The above rule is not applicable to persons entering the
" private " wards.

10. Two beds (one in the men's and one in the women's
wards) shall be kept in reserve to receive such patients as
may, from sudden accident, be in need of immediate relief.

11. The applicants for admission shall attend at the place
appointed for that purpose, in the Hospital, at any time

from eleven o'clock A.M. till one o'clock P.M., and shall present the following certificate to the attending Medical Officer:—

Form of Certificate of the admission of " Ordinary " Patients.

MONTREAL, day of ,187
Sir,—I recommend a resident of this City, to be admitted into the Montreal General Hospital as an " Ordinary " Patient if be thought a fit object for admission.
To the attending Medical Officer of the
Montreal General Hospital.

Form of Certificate for the admission of Pay Patients.

MONTREAL, day of , 187
Sir,—I recommend , who has resided in the parish of for to be admitted into the Montreal General Hospital as a pay-patient, if be thought a fit object for admission, and I undertake personally to pay to the Hospital the sum or sums which shall become due on such admission.
To the attending Medical Officer of the
Montreal General Hospital.

CHAPTER XIII.

OF PATIENTS AND VISITORS.

1st. The object of admitting patients into the Hospital being their relief from bodily ailment, the first consideration must be to treat them medically, and to avoid therefore all needless excitement as detrimental to their cure.

2nd. That every patient admitted shall be requested to state his or her name, sex, age, religion, birthplace, residence and disease.

3rd. It shall be the duty of the House Surgeon to obtain

these particulars, and to have them properly entered upon the usual card and have it affixed to the head of the bed in which the patient is placed.

This record to be treated as conclusive unless and until the patient voluntarily requests the House Surgeon to make any changes in the card, such request to be made in the presence of at least one other resident official of the Hospital, as a witness thereto.

The record of the religion on the card to be an invitation to clergymen of the same denomination as the patient to attend on him ; and the officials of the Hospital are forbidden to permit any interference with the conditions recorded on the card for the time being.

4th. A daily list of the patients in the Hospital, containing the records in clause two, to be exhibited in the hall for the information of clergymen of different denominations, visiting the Hospital.

5th. Visitors to be admitted to see patients under such regulations as may be made from time to time by the Committee of Management, and whilst there to attend to the directions of the House Surgeon, who will be held responsible that no conduct is permitted on the part of visitors detrimental to the patients, or to interfere with their proper medical treatment and care.

6th. The names of all visitors to be entered in a book to be kept for that purpose in the entrance hall.

CHAPTER XIV.

OF VACCINATION.

1. On any of the days appointed by the Medical Board for vaccinating the poor, gratis, any person or persons, without the recommendation required for the admission of patients to the Hospital, may present their children for vaccination.

2. The person or persons who shall vaccinate any children or other persons, on any of the above stated days, shall

register in a book belonging to the Hospital, kept for that purpose, the name, age, sex, and religion of each and every individual so vaccinated; the part of the body on which the vaccination has been performed; the date, progress, and termination of each and every case.

CHAPTER XV.

OF THE PENALTY OF RECEIVING ANY FEE OR REWARD.

1. No Governor, officer, or other person connected with the Montreal General Hospital, shall at any time presume, under any pretence whatever, on pain of expulsion, to give to, or take from any tradesman, patient, servant, or any other person or persons, any fee, reward or gratuity of whatever kind, directly or indirectly, for any services done or to be done, or proposed to be done, on account of or relative to the said Hospital, except in the case of Physicians to the Institution, who, for the accommodation of their private patients, may bring them into the private wards when it does not interfere with the objects of the Institution.

CHAPTER XVI.

When an ordinary patient dies in the Hospital (if the body is to be interred), the Steward shall give notice thereof to the clergyman to whose congregation or church he or she may belong; and the Porter shall, if necessary, assist the Sexton at the interment. Immigrant paupers of less than twelve months' residence in the country, who die in the Hospital, shall have their funerals defrayed from the money received from the Immigrant funds. With regard to other ordinary patients, the Committee of Management are authorised to request the different religious communities to which they may have belonged, to provide for the funerals of such unless the bodies are otherwise disposed of by law.

CHAPTER XVII.

OF ALTERING AND REPEALING BY-LAWS.

No additional By-Law shall hereafter be made, nor shall any of those already existing be in any way altered or repealed except a special motion for such alteration, repeal, &c., be made and received at one of the quarterly meetings of the Governors, and passed at the next quarterly meeting, before being submitted to the Corporation.

REGULATIONS CONCERNING VISITORS.

1. Patients in the Public Wards will be allowed to see their friends on Tuesday, Thursday and Saturday, between 3 and 4 o'clock, p.m.—subject to restrictions by the House Surgeon as to number and duration of visits.

The House Surgeon has discretionary power to admit visitors to any patient at any time on special passes to be issued by him or, in his absence, by the Assistant House Surgeon.

2. Ministers of Religion and other persons, visiting the Hospital with the sanction of the Committee of Management, for the purpose of religious ministrations, will be admitted every day between 2 and 5.30 p.m. In cases of urgency the House Surgeon may send for a Clergyman at any hour, when requested by a patient.

3. The House Surgeon or, in his absence, the Assistant House Surgeon, shall have the power to prevent visitors from having access to any patient, even during the hours hereby fixed for that purpose, when special circumstances render it expedient so to do. But in such cases he shall report his action to the next meeting of the Committee.

4. Patients in Private Wards may be visited by their friends at any suitable hour in the day time, with the permission of the House Surgeon or Lady Superintendent, and not otherwise. No visitor to remain after 8 p.m.

5. All visitors will be required to enter their Names on the occasion of every visit.

By order of the Committee of Management.

MONTREAL, April 10, 1876.

CONTENTS.

— :o: — —

www.ingramcontent.com/pod-product-compliance
Lightning Source LLC
Chambersburg PA
CBHW022031190326
41519CB00010B/1669